STEM IS EVERYWHERE!

SMARTPHONES

COMPUTERS IN YOUR POCKET

EMMETT MARTIN

Gareth Stevens PUBLISHING

Please visit our website, www.garethstevens.com.
For a free color catalog of all our high-quality books,
call toll free 1-800-542-2595 or fax 1-877-542-2596.

Portions of this work were originally authored by Alicia Z. Klepeis and published as *How Smartphones Work* (Everyday STEM). All new material in this edition authored by Emmett Martin.

Library of Congress Cataloging-in-Publication Data
Names: Martin, Emmett, author.
Title: Smartphones : computers in your pocket / Emmett Martin.
Description: New York : Gareth Stevens Publishing, [2023] | Series: Stem is everywhere! | Includes bibliographical references and index.
Identifiers: LCCN 2022024469 (print) | LCCN 2022024470 (ebook) | ISBN 9781538283615 (library binding) | ISBN 9781538283592 (paperback) | ISBN 9781538283622 (ebook)
Subjects: LCSH: Smartphones–Juvenile literature.
Classification: LCC TK6564.4.C45 M37 2023 (print) | LCC TK6564.4.C45 (ebook) | DDC 621.3845/6–dc23/eng/20220708
LC record available at https://lccn.loc.gov/2022024469
LC ebook record available at https://lccn.loc.gov/2022024470

Published in 2023 by
Gareth Stevens Publishing
2544 Clinton Street
Buffalo, NY 14224

Designer: Tanya Dellaccio
Editor: Therese Shea

Photo credits: Series Art Supphachai Salaeman/Shutterstock.com; Cover sdecoret/Shutterstock.com; p. 4 APISITH/Shutterstock.com; p. 5 Rido/Shuttertock.com; p. 7 (top) POM POM/Shutterstock.com; p. 7 (bottom) https://upload.wikimedia.org/wikipedia/commons/1/16/Alexander_Graham_Telephone_in_Newyork.jpg; p. 9 Jan Jirous/Shutterstock.com; p. 11 Cary Kalscheuer/Shutterstock.com; p. 13 Poravute Siriphiroon/Shutterstock.com; p. 15 (top) A_B_C/Shutterstock.com; p. 15 (bottom) Alexey Boldin/Shutterstock.com; p. 16 Kamil Urban/Shutterstock.com; p. 17 McLittle Stock/Shutterstock.com; p. 18 https://upload.wikimedia.org/wikipedia/commons/9/9e/IBM_Simon_Personal_Communicator.png; p. 19 Chris Willson/Alamy Images; p. 20 https://upload.wikimedia.org/wikipedia/commons/f/ff/Palm-IMG_7025.jpg; p. 21 Allmy/Shutterstock.com; p. 22 https://upload.wikimedia.org/wikipedia/commons/4/4e/Eniac.jpg; p. 23 Stock Foto.Touch/Shutterstock.com; p. 25 (left) ZernLiew/iStock.com; p. 25 (right) Remus Rigo/Shutterstock.com; p. 27 adriaticfoto/Shutterstock.com.

Printed in the United States of America

CPSIA compliance information: Batch #CWGS23: For further information contact Gareth Stevems Publishing at 1-800-398-2504.

CONTENTS

Words in the glossary appear in **bold** type the first time they are used in the text.

SMARTPHONE WORLD

It's 7:00 a.m. An alarm buzzes. A teenager picks up his smartphone and silences the buzzing. The screen recognizes his face and opens the phone. He checks an **app** he uses to chat with friends. He gets up, dresses, and has breakfast. As he's waiting for the bus, he takes out his smartphone. He plays a game and puts the phone away when the bus arrives.

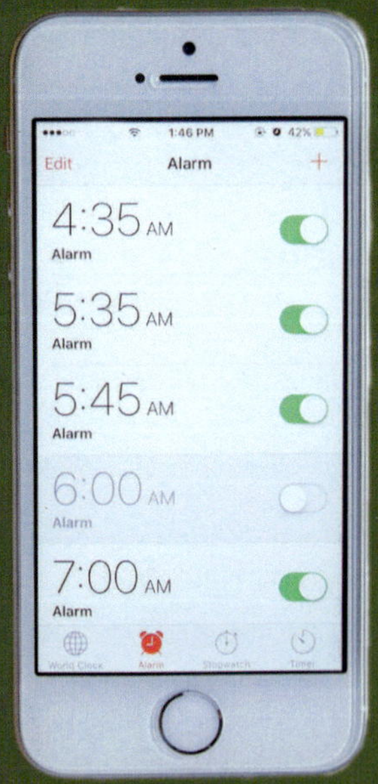

In chemistry class, a teacher asks if anyone can record an experiment. He raises his hand and takes out his phone. That night, he sets his phone alarm for the next morning. For many, smartphones are a part of life!

ACCORDING TO A 2019 STUDY, ABOUT 53 PERCENT OF AMERICAN CHILDREN HAVE SMARTPHONES BY THE TIME THEY'RE 11 YEARS OLD.

WHAT MAKES A PHONE "SMART"?

A SMARTPHONE IS A MIX OF A TELEPHONE AND A COMPUTER THAT YOU CAN TAKE WITH YOU WHEREVER YOU GO. THE COMPUTER PART IS WHAT MAKES THE PHONE "SMART." IT CAN HOLD MANY APPS (APPLICATIONS), INCLUDING INTERNET **BROWSERS**. SMARTPHONES USUALLY HAVE A CAMERA AND A TOUCH SCREEN FOR USERS TO INPUT COMMANDS.

FROM WIRED TO WIRELESS

To understand smartphones and how they work, let's learn about earlier phones. In 1876, Alexander Graham Bell patented an invention that we call the telephone. It changed the sound waves of a caller's voice into electrical signals. The signals traveled through wires to a receiving telephone, which could change the signals back into the sound waves of the caller's voice.

A process invented later allowed the sound waves to travel as light waves through glass fibers, a **technology** called fiber optics. Radio waves transmitted by antennae and **satellites** are yet another way to carry signals to a receiving phone.

CUTTING THE CORD

A LOT HAS CHANGED SINCE BELL'S TELEPHONE! CELLULAR (OR CELL) PHONES INTRODUCED WIRELESS COMMUNICATION. THE FIRST CELL PHONES ONLY ALLOWED PEOPLE TO TALK TO EACH OTHER. THEY DIDN'T HAVE THE SPECIAL FEATURES SMARTPHONES HAVE TODAY. BUT EARLY CELL PHONES CAME IN MANY SIZES, SHAPES, AND COLORS, LIKE SMARTPHONES DO.

IN THIS PHOTO, ALEXANDER GRAHAM BELL USES A TELEPHONE IN 1892, THE YEAR TELEPHONE LINES CONNECTED NEW YORK CITY AND CHICAGO, ILLINOIS, FOR THE FIRST TIME.

Let's fast-forward to today, the age of the smartphone. Smartphones are a bit like small radios. Like radios, they use radio-wave technology to receive sounds and send them out too. When someone makes a call on a smartphone, the phone sends radio signals out from its antenna. These invisible signals travel to an antenna at the closest cell tower, or base station.

FROM SOUND WAVE TO RADIO WAVE

A SMARTPHONE CONTAINS A TRANSMITTER, FOR SENDING SIGNALS, AND A RECEIVER, FOR RECEIVING SIGNALS. A MICROPHONE CHANGES YOUR VOICE'S SOUND WAVES INTO ELECTRICAL SIGNALS. A MICROCHIP TURNS THESE SIGNALS INTO A RADIO WAVE, WHICH IS SENT OUT THROUGH THE ANTENNA TO A CELL TOWER. FROM THERE, IT'S SENT TO THE RECEIVING PHONE WHERE THE PROCESS IS REVERSED SO THE VOICE CAN BE HEARD.

The tower receives the radio signal. It sends the signal to the receiving phone. Depending on where the receiver of the call is located, the signal might be transferred, or relocated, to a different base station.

THE CIRCULAR AND RECTANGULAR SHAPES ON CELL TOWERS ARE ANTENNAE THAT TRANSMIT AND RECEIVE SIGNALS.

NETWORKING

A smartphone needs a connection to a cell phone network. Cell phone networks are divided into areas known as cells. A cell is simply the area surrounding a base station.

A city or town is divided into many cells. Each cell has a tower. Every tower has electronic equipment, or gear, and antennae. This equipment can both receive and send out radio signals. Towers may be attached to tall structures already in place, such as bridges and streetlights.

As smartphone users move around, their phones switch cells. This allows people to keep talking even while moving from place to place, without the call ending.

SOMETIMES CELL TOWERS ARE MADE TO LOOK LIKE SOMETHING ELSE, LIKE A TREE, SO IT FITS INTO ITS SURROUNDINGS BETTER.

MORE ABOUT TOWERS

ONE TOWER MIGHT HANDLE THOUSANDS OF CALLS AT ONCE. THE HEIGHT OF A CELL TOWER AFFECTS ITS RANGE. THE TALLEST ONES MAY BE 400 FEET (122 M) TALL AND SEND AND RECEIVE SIGNALS FROM AS FAR AS 45 MILES (72 KM) AWAY. SOME TOWERS ONLY SERVE A SMALLER AREA, SUCH AS AN AIRPORT.

HARDWARE AND SOFTWARE

While a smartphone is like a radio, it's not just a radio. A smartphone is also a little computer. In fact, it has all the same parts as a full-sized computer.

The operating system (OS) is one of the most important features on a smartphone. It's what runs the hardware and software on the phone. Hardware is all the wiring and other **physical** parts you could see if you opened a smartphone to look at its insides. Hardware includes the battery that provides power. It includes storage devices that help save information, or data. It includes the camera too.

A LOT OF PIECES OF HARDWARE ARE NEEDED TO MAKE A SMARTPHONE WORK.

METAL MATTERS

SMARTPHONES CONTAIN DIFFERENT METALS, INCLUDING ALUMINUM AND LITHIUM. GOLD, SILVER, AND PLATINUM MAY ALSO BE FOUND INSIDE. A SMARTPHONE ALSO HAS WHAT ARE SOMETIMES CALLED RARE-EARTH ELEMENTS. MOST PEOPLE HAVE NEVER HEARD OF YTTRIUM OR TERBIUM, YET THESE ELEMENTS ARE NECESSARY IN SMARTPHONES. MANY OF THESE METALS ARE HARD TO REMOVE FROM THE EARTH.

Have you ever received a call while you were watching a video on a smartphone? Your phone went back to the video when the call was done, right? The OS manages the software that makes it all possible.

Software is all the programs that tell the smartphone what to do. This includes apps. The OS allows the smartphone to run different apps at the same time. Software makes the computer inside the smartphone easy to use.

Different brands of smartphones may have different operating systems. For example, iPhones use iOS. Samsung, LG, Google, and many other smartphones use Android OS.

PLASTIC

MANY SMARTPHONE CASINGS ARE MADE OF PLASTIC. PLASTIC IS AN INEXPENSIVE MATERIAL BUT CAN ALSO BE STRONG. PLASTIC IS IN THE LCD (LIQUID CRYSTAL DISPLAY) TOUCH SCREEN. THIS IS A SCREEN FOUND ON MANY ELECTRONICS, SUCH AS DIGITAL WATCHES, TVs, AND CELL PHONES, THAT WORKS BY PASSING A SMALL ELECTRIC CURRENT THROUGH A SPECIAL LIQUID.

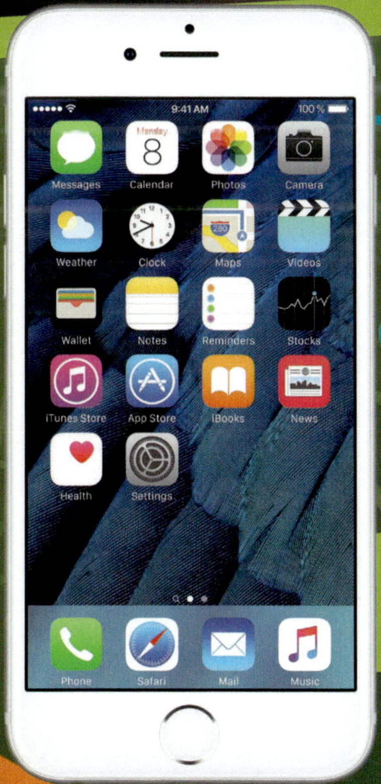

APPS HELP YOU TELL TIME, SEND MESSAGES, CHECK EMAIL, AND DO MANY OTHER TASKS ON A SMARTPHONE.

WHAT'S WI-FI?

When people are at home, they often connect their smartphones to their Wi-Fi, or a wireless internet connection. A Wi-Fi network requires a router, which is a device that transmits the internet signal as radio waves to any device equipped for Wi-Fi. A Wi-Fi signal for a network in someone's home usually only travels 150 feet (46 m) indoors or about 300 feet (92 m) outdoors.

A FIRST IN WI-FI

GRAND HAVEN, MICHIGAN, BECAME THE FIRST CITY IN THE UNITED STATES TO HAVE CITYWIDE WI-FI IN 2004. PEOPLE CAN CONNECT WIRELESSLY TO THE INTERNET OVER THE CITY'S ENTIRE 6 SQUARE MILES (15.5 SQ KM)—AND ALSO 15 MILES (24 KM) INTO LAKE MICHIGAN. MAINTAINING A HOME INTERNET CONNECTION COSTS MONEY SO CITY-WIDE WI-FI HELPS PEOPLE SAVE MONEY.

WI-FI SIGNALS WEAKEN WHEN THEY GO THROUGH WALLS OR FLOORS. SMARTPHONES HAVE A WI-FI SIGNAL INDICATOR. THE MORE CURVED BARS THAT ARE FILLED IN, THE STRONGER THE SIGNAL.

When people aren't at home, such as at a café, people may have to enter a password to connect to a place's Wi-Fi. Using Wi-Fi to connect to the internet drains less battery power than using a cellular network.

CREATING THE SMARTPHONE

We think of smartphones as speedy, lightweight, and multitasking. Today's smartphones are all these things, but the smartphones of the past weren't. **Developing** a truly mobile, or moveable, and useful smartphone was challenging.

MEET SIMON

THE IBM COMPANY MADE THE FIRST SMARTPHONE IN 1992. THE PHONE, NAMED THE SIMON PERSONAL COMMUNICATOR, HAD A TOUCHSCREEN AND SOFTWARE FOR CHECKING EMAIL. THIS EARLY SMARTPHONE ALSO HAD A CALCULATOR, ADDRESS BOOK, AND CALENDAR. BUT SIMON WAS BIG AND HEAVY, WEIGHING OVER 1 POUND (0.45 KG). AND IT WAS $1,100—PRETTY COSTLY FOR THE TIME.

In April 1973, an employee of the Motorola company named Martin Cooper made the world's first cell phone call. More than 10 years later, the Motorola DynaTAC 8000x became available to the public. It cost about $4,000 and weighed about 2 pounds (907 g). Some called it "the Brick." After only about 30 minutes of talk time, this phone's battery died. And it took 10 hours to recharge!

THE MOTOROLA DYNATAC WAS THE FIRST CELL PHONE AVAILABLE TO THE PUBLIC TO BUY.

Let's move on to the 1990s. Though the public now used cell phones, the phones were only used to make calls. What about people who wanted a portable device for storing information? They used personal digital assistants (PDAs). The Palm Pilot was a popular PDA model. It had a touch screen. It also had memory for storing information and could be used with a computer.

Newer cell phones and PDAs began to include email and message apps. Wanting to make an even better tool, companies combined the technologies of cell phones and PDAs. The result was the first smartphone.

THE PALM PILOT WAS SMALL ENOUGH TO FIT IN SOMEONE'S POCKET, WHICH MADE IT EASY TO CARRY AROUND.

IMPROVEMENTS

MUCH HAS IMPROVED SINCE THE EARLY DAYS OF SMARTPHONES. THEIR BATTERIES LAST LONGER. THEIR TOUCH SCREENS ARE EASY TO USE. THEIR CAMERAS TAKE EXCELLENT PHOTOS. TODAY'S SMARTPHONES COME WITH GPS (GLOBAL POSITIONING SYSTEM) TECHNOLOGY. THIS ALLOWS US TO EASILY GET DIRECTIONS AND LET PEOPLE KNOW WHERE WE ARE.

FROM BIG TO SMALL

The first computers were so large they filled entire rooms! Today's smartphones are filled with information and processing power, yet they fit in the palm of a person's hand. Over time, engineers and scientists were able to make computer parts small but powerful. Technology is still progressing in this direction.

THIS IS A PHOTO OF THE PROGRAMMABLE ELECTRONIC DIGITAL COMPUTER CALLED ENIAC (ELECTRONIC NUMERICAL INTEGRATOR AND COMPUTER) FROM THE 1940s.

NANO SIM CARD

THE SIM CARD IN A SMARTPHONE TELLS NETWORKS THE DEVICE'S PHONE NUMBER AND OTHER IDENTIFYING FEATURES. (SIM STANDS FOR "SUBSCRIBER IDENTITY MODULE.") THE LATEST DEVELOPMENT FOR THIS PART IS THE NANO SIM CARD. IT'S MUCH SMALLER THAN THE SIM CARDS THAT CAME BEFORE IT, ALLOWING MORE SPACE FOR OTHER PHONE PARTS INSIDE THE DEVICE.

Nanotechnology is a field of science focused on studying and making super-small things. How small? A nanometer is one-billionth of a meter. If that still isn't clear, a sheet of paper is around 100,000 nanometers thick! Nanotechnology has led to many developments in smartphones, allowing them to be thinner, lighter, and more durable, or sturdy and lasting, than ever before.

THE CPU

The central processing unit (CPU) is the "brain" of a smartphone, containing the programming for the OS. For most smartphones today, the CPU is located on something called a system-on-a-chip (SoC). This chip is tiny, yet it contains essential parts to make the phone work, including the parts that store data and the **modem** that helps connect the phone to wireless signals.

Instead of having many separate parts that take up space, the SoC holds all of them in one place. Other physical parts, such as **RAM** and the camera, connect to the SoC. Without these small chips, smartphones would have to be much bigger.

THIS IMAGE SHOWS THE LAYERS OF A SMARTPHONE POWERED BY A BATTERY. THE MOTHERBOARD HOLDS THE SOC.

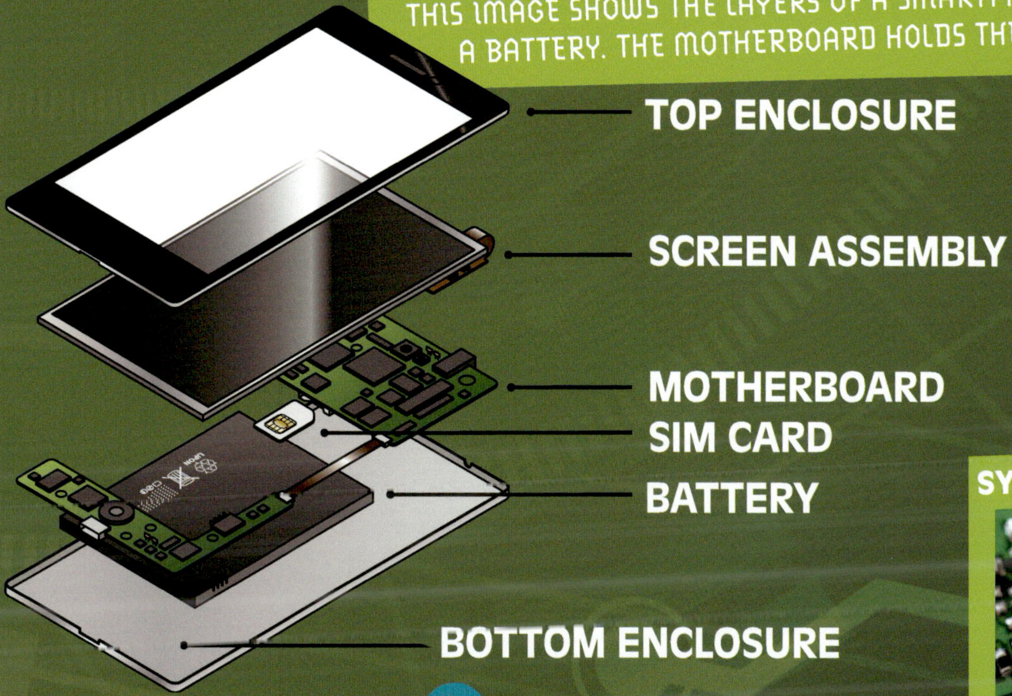

SYSTEM-ON-A-CHIP

HIGH-SPEED PROCESSING

THE IPHONE 13 CAN PROCESS 15.8 TRILLION OPERATIONS, OR INSTRUCTIONS, PER SECOND. (THAT'S 15,800,000,000,000 OPERATIONS!) IT HAS AN A15 BIONIC SYSTEM-ON-A-CHIP CONTAINING 15 BILLION (15,000,000,000) TRANSISTORS, WHICH ARE DEVICES FOR CONTROLLING ELECTRICAL SIGNALS. SOCs WILL ONLY GET MORE POWERFUL AS NANOTECHNOLOGY CONTINUES TO DEVELOP.

HOW SMARTPHONES CAN HELP

We use smartphones mostly for calling friends, finding our way in new areas, watching videos, texting with others, and playing games. But smartphones are helping people in new ways all the time. For example, certain apps can help us **translate** languages to communicate with people who speak languages we don't understand. They've also become aids to people with disabilities.

Blind or partially sighted users can use an app, such as Lookout, that will tell them what the words say on road signs, menus, or other locations. Another function within the Google Maps app can inform users of the street they're walking on and when to make a turn toward where they're going.

TECHNOLOGY COMPANIES ARE HOPING THAT SMARTPHONES WILL BE ABLE TO RESPOND TO GESTURES AS COMMANDS ONE DAY. THIS COULD BE HELPFUL FOR THOSE WHO COMMUNICATE THROUGH SIGN LANGUAGES.

QUICK CALL FOR ASSISTANCE

THE BE MY EYES APP CONNECTS PEOPLE WHO ARE VISUALLY IMPAIRED OR BLIND WITH SIGHTED HELPERS ALL OVER THE WORLD. A PERSON REQUESTS ASSISTANCE THROUGH THE APP. A **VOLUNTEER** HELPER ACCEPTS THE REQUEST. THROUGH A VIDEO CALL, THE VOLUNTEER CAN HELP WITH WHATEVER TASK IS NEEDED, SUCH AS READING A SIGN.

FUTURE DEVELOPMENTS

Smartphones have gotten better over time. Their cameras have become more powerful, their operations faster, their memories larger, and their batteries longer lasting. Smartphones are becoming slightly larger because people enjoy viewing a larger screen. Some companies are making larger phones that can fold to fit into the user's pocket.

What will smartphones be like in ten years? Some people would like to keep their phones longer, swapping out old hardware for new. Some think smartphones might be replaced by something that looks like a pair of glasses. It will be up to future engineers and computer scientists to figure out how to do this!

A TIMELINE OF SMARTPHONE TECHNOLOGY

1876 ALEXANDER GRAHAM BELL RECEIVES A PATENT FOR HIS TELEPHONE.

1973 THE WORLD'S FIRST CELLULAR PHONE CALL IS MADE.

1974 A WATCH IS THE FIRST PRODUCT TO FEATURE A SYSTEM-ON-A-CHIP.

1984 THE MOTOROLA DYNATAC 8000X CELL PHONE GOES ON SALE.

1992 IBM CREATES THE SIMON PERSONAL COMMUNICATOR, THE WORLD'S FIRST SMARTPHONE.

2007 THE APPLE IPHONE IS RELEASED.

2008 THE FIRST ANDROID PHONE, THE T-MOBILE G1, IS LAUNCHED.

2010 THE MOTOROLA ATRIX IS THE FIRST PHONE TO USE A FINGERPRINT RECOGNITION SYSTEM.

2016 THE IPHONE 7 PLUS IS THE FIRST WATERPROOF IPHONE.

2017 FACE RECOGNITION TECHNOLOGY ALLOWS USERS TO UNLOCK SMARTPHONES BY LOOKING AT THEM.

2018 A SMARTPHONE CALLED THE ROYOLE FLEXPAI BECOMES THE FIRST FOLDING PHONE OF ITS KIND.

2021 ABOUT 3.95 BILLION PEOPLE, OR HALF THE WORLD'S POPULATION, OWN A SMARTPHONE.

SMARTPHONE COMPANIES HAVE MADE GREAT TECHNOLOGICAL PROGRESS WITH EACH NEW FORM OF THEIR PRODUCT.

GLOSSARY

app: Short for application. Computer and smartphone programs that are made to perform certain functions, or tasks.

browser: A computer application for locating, looking at, and interacting with websites.

develop: To put effort into producing something over time.

GPS: Stands for Global Positioning System. A system that uses satellite signals to locate places on Earth.

modem: A machine that changes signals produced by one type of device into a form that can be used with another type of device.

physical: Having a form that can be seen or touched.

RAM: Stands for random-access memory. A computer in which data is stored for a limited amount of time as a program is in use.

satellite: A machine sent into space that moves around Earth, the moon, the sun, or a planet to complete scientific research or a type of communication.

technology: Using science, engineering, and other industries to invent useful tools or to solve problems. Also a machine, piece of equipment, or method created by technology.

translate: To tell what words in one language mean in another.

volunteer: A person who offers service without being paid and sometimes without being asked.

FOR MORE INFORMATION

BOOKS

Dean, Mary. *Smartphones.* Hollywood, FL: Mason Crest, 2022.

Gitlin. Marty. *Smartphones.* Ann Arbor, MI: Cherry Lake Publishing, 2019.

Kelly, Tracey. *Communication Technology: From Smoke Signals to Smartphones.* Tucson, AZ: Brown Bear Books, 2019.

WEBSITES

How Cell Phones Work
www.explainthatstuff.com/cellphones.html
Do you need more in-depth information about phone tech? Check out this page.

How Do Cell Towers Work?
wonderopolis.org/wonder/how-do-cell-towers-work
Find out more facts about this essential part of smartphone technology.

Telephone
kids.britannica.com/kids/article/telephone/353842
Read more about how what you say into the phone travels to someone else's phone.

Publisher's note to educators and parents: Our editors have carefully reviewed these websites to ensure that they are suitable for students. Many websites change frequently, however, and we cannot guarantee that a site's future contents will continue to meet our high standards of quality and educational value. Be advised that students should be closely supervised whenever they access the internet.

INDEX